POISONOUS CREATURES

INVESTIGATE!

Beware the PUFFER FISH!

Howard Phillips

Enslow PUBLISHING

Please visit our website, www.enslow.com. For a free color catalog of all our high-quality books, call toll free 1-800-398-2504 or fax 1-877-980-4454.

Library of Congress Cataloging-in-Publication Data

Names: Phillips, Howard, 1971- author.
Title: Beware the puffer fish / Howard Phillips.
Description: New York : Enslow Publishing, [2023] | Series: Poisonous creatures | Includes index.
Identifiers: LCCN 2021050309 (print) | LCCN 2021050310 (ebook) | ISBN 9781978527522 (library binding) | ISBN 9781978527508 (paperback) | ISBN 9781978527515 (6 pack) | ISBN 9781978527539 (ebook)
Subjects: LCSH: Puffers (Fish)–Juvenile literature. | Poisonous animals–Juvenile literature.
Classification: LCC QL638.T32 P53 2023 (print) | LCC QL638.T32 (ebook) | DDC 597/.64–dc23/eng/20211013
LC record available at https://lccn.loc.gov/2021050309
LC ebook record available at https://lccn.loc.gov/2021050310

Portions of this work were originally authored by Alicia Z. Klepeis and published as *The Puffer Fish*. All new material in this edition was authored by Howard Phillips.

Published in 2023 by
Enslow Publishing
29 E. 21st Street
New York, NY 10010

Designer: Katelyn E. Reynolds
Interior Layout: Tanya Dellaccio
Editor: Greg Roza

Photo credits: Series background Vector Tradition/Shutterstock.com; cover (puffer fish) jara3000/Shutterstock.com; pp. 5 (top), 9 (top) KT photo/Shutterstock.com; p. 5 (bottom) Arunee Rodloy/Shutterstock.com; p. 6 Toxotes Hun-Gabor Horvath/Shutterstock.com; p. 7 (inset) Tatiana Belova/Shutterstock.com; p. 7 (main) Rich Carey/Shutterstock.com; p. 9 (bottom) axell.rf/Shutterstock.com; p. 10 Richard Whitcombe/Shutterstock.com; p. 11 Eric Isselee/Shutterstock.com; p. 12 robertzwinchell/Shutterstock.com; p. 13 (inset) Photos by D/Shutterstock.com; p. 13 (main) Kurit afshen/Shutterstock.com; p. 15 (inset) Jnichanan/Shutterstock.com; p. 15 (main) Strongstockphotos/Shutterstock.com; p. 18 CK Ma/Shutterstock.com; p. 19 Franny Constantina/Shutterrstock.com; p. 20 Brent Barnes/Shutterstock.com; p. 21 Toxotes Hun-Gabor Horvath/Shutterstock.com; p. 23 (inset) Kerry L. Werry/Shutterstock.com; p. 23 (main) Rich Carey/Shutterstock.com; p. 25 (inset) HikoPhotography/Shutterstock.com; p. 25 (main) medo.gw78/Shutterstock.com; p. 26 (right) Nantawat Chotsuwan/Shutterstock.com; p. 26 (left) slowmotiongli/Shutterstock.com; p. 27 Aries Sutanto/Shutterstock.com; p. 29 (main) mattjamesphotography/Shutterstock.com; p. 29 (inset) https://upload.wikimedia.org/wikipedia/commons/9/92/Tetraodon_pustulatus_-_Cross_River_Puffer_%288767208722%29.jpg.

Printed in the United States of America

CPSIA compliance information: Batch #CSENS23: For further information contact Enslow Publishing, New York, New York, at 1-800-398-2504.

CONTENTS

WORDS IN THE GLOSSARY APPEAR IN **BOLD** TYPE THE FIRST TIME THEY ARE USED IN THE TEXT.

SMALL, CUTE,... AND DEADLY

Puffer fish are unusual marine animals. They usually have lean bodies and round, bulging heads. In times of danger, however, they look like living beach balls! These fish may appear harmless, and some are even really cute. But look out—many puffer fish are armed with deadly poison.

In September 1774, explorer Captain James Cook recorded in his log the effects of puffer fish poison. Members of his crew had numbness and shortness of breath after eating the fish. Some pigs on board ate the remains of the puffers and died.

SOME PUFFER FISH
HAVE SPIKES. THEY ARE
SOMETIMES CALLED
PORCUPINE FISH.
Get the Facts!
People disagree about where and when it was discovered that puffer fish were toxic. Written records discussing puffer fish from China and Egypt are thousands of years old.

ALL KINDS OF PUFFERS

Most puffers live in **tropical** and subtropical ocean waters. They are found around the world in the Atlantic, Pacific, and Indian Oceans. Puffer fish have almost never been seen in cold water. About 29 species, or kinds, of puffers live in fresh water. Others survive in **brackish** water.

Get the Facts!

Some puffers don't have bright coloring. They may blend in with their environment. This can help them avoid predators.

Some puffer fish are tiny. The dwarf, or pygmy, puffer can be less than 1 inch (2.5 cm) long! Others are big. The giant puffer can be more than 2 feet (61 cm) long. Some puffer fish are brightly colored. This coloring may warn predators to stay away.

NO EASY MEAL

Puffers are slow, clumsy swimmers, which would make them an easy target for many predators. However, many puffer fish have an amazing **defense**. When threatened, or feel in danger, they will take in large amounts of water (and sometimes air). This quickly **inflates** their bodies to two or three times their normal size.

Some puffers also have very rough skin. Others have spikes that stick out of their skin when inflated. This makes them look like a spiky balloon! These defenses make it hard for an enemy to eat the puffer fish.

THE PUFFER FISH SHOWN HERE WAS SPOTTED NEAR A CORAL REEF IN THE ANDAMAN SEA NEAR THAILAND.

Get the Facts!

Tiger sharks, sea snakes, and lizard fish can successfully eat puffers without getting harmed. They are **immune** to puffer poison.

PUFFER FISH PARTS

Puffer fish come in many shapes, sizes, and colors. Some have spikes, and some do not. Many are poisonous, but some are not. All puffers are capable of puffing up! Other than the spikes, all puffers share the features shown on page 11.

DIODON HOLOCANTHUS, SHOWN HERE, HAS MANY DIFFERENT COMMON NAMES, INCLUDING BALLOON FISH, BLOTCHED PORCUPINE FISH, FRECKLED PORCUPINE FISH, HEDGEHOG FISH, AND SPINY PUFFER.

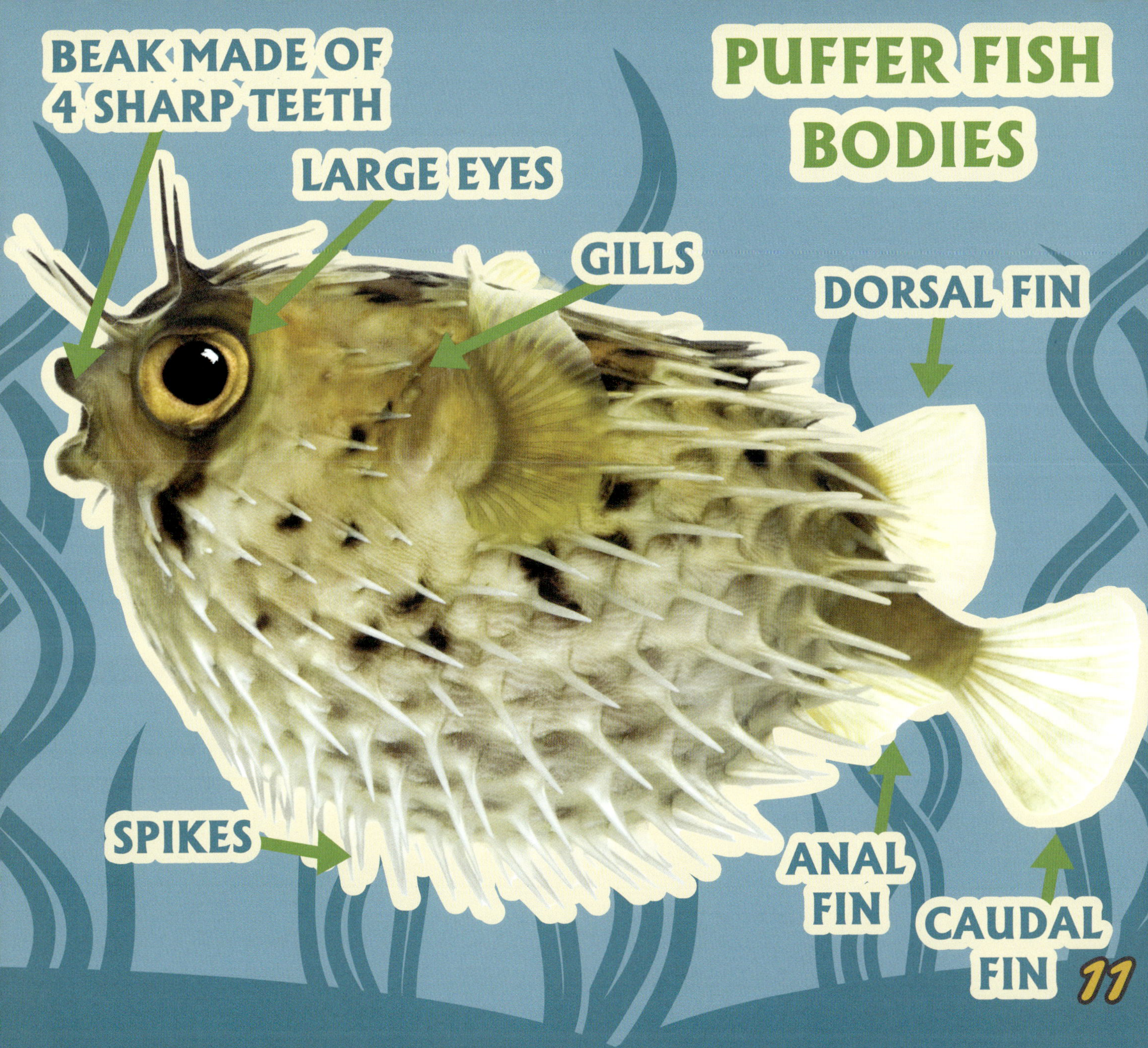
PUFFER FISH BODIES
BEAK MADE OF 4 SHARP TEETH
LARGE EYES
GILLS
DORSAL FIN
SPIKES
ANAL FIN
CAUDAL FIN

FINDING FOOD

Puffer fish don't normally make a surprise attack on **prey** from a hidden location. Instead, they feed on slow-moving or stationary animals. Puffer fish look for food by probing the sea floor. They also check the nooks and crannies found in coral reefs. Puffers have big, sensitive lips. These allow them to tell prey apart from objects such as rocks.

Puffer fish are omnivores, which means they eat plants and animals. Their diet includes fish, **algae**, worms, sponges, and more.

PUFFER FISH BEAKS ARE VERY SHARP, AND THEIR BITE IS VERY STRONG. THEY HAVE TAKEN OFF THE FINGERTIPS OF FISHERMEN WHO WERE REMOVING PUFFERS FROM THEIR HOOKS.

Puffers have four teeth that form a beak-like structure. Their beak is so strong it can crush the shells of oysters and clams! This wears down their razor-sharp teeth, which grow throughout their lives.

YOUNG PUFFERS

Puffer fish lay eggs. Each female will lay three to seven eggs at a time. The hard shell of the eggs keeps baby puffer fish safe until they hatch.

Young puffers are not as poisonous as their parents right away. They have not built up their stores of poison yet. Luckily, puffer moms give some poison to their young. Scientists have found this poison in the eggs and on the surface of puffer fish larvae. It might not be enough to kill a predator. But, some predators that try to eat puffer fish larvae spit them out.

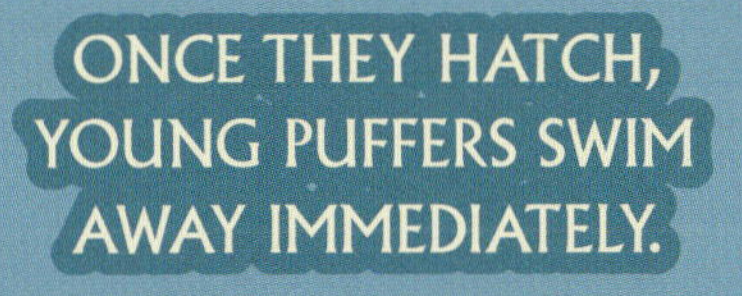

The life-span of a puffer fish is typically four to eight years, depending on the species. Puffers kept in aquariums may live 10 to 12 years.

GETTING TO KNOW THE PUFFER FISH

Scientific Name

Puffer fish are in the animal family Tetraodontidae.

Common Name

Puffer fish, blowfish, globefish, balloon fish, porcupine fish. In Japan, puffer fish prepared as a dish is called fugu. This comes from the Chinese characters for "river" and "pig."

Range

Most commonly found in tropical and subtropical waters of the Indian, Pacific, and Atlantic Oceans. They are also found in some fresh water and brackish waters.

Puffers vary widely in size, from less than 1 inch (2.5 cm) to about 3 feet (0.9 meters) long.

How Many Kinds

There are more than 120 species of puffer fish.

Puffer Fish as Food

Fugu is considered a delicacy in Japan, but it's illegal in other parts of the world. It's illegal in Japan to serve puffer fish to the emperor because one puffer fish has enough poison to kill 30 adults!

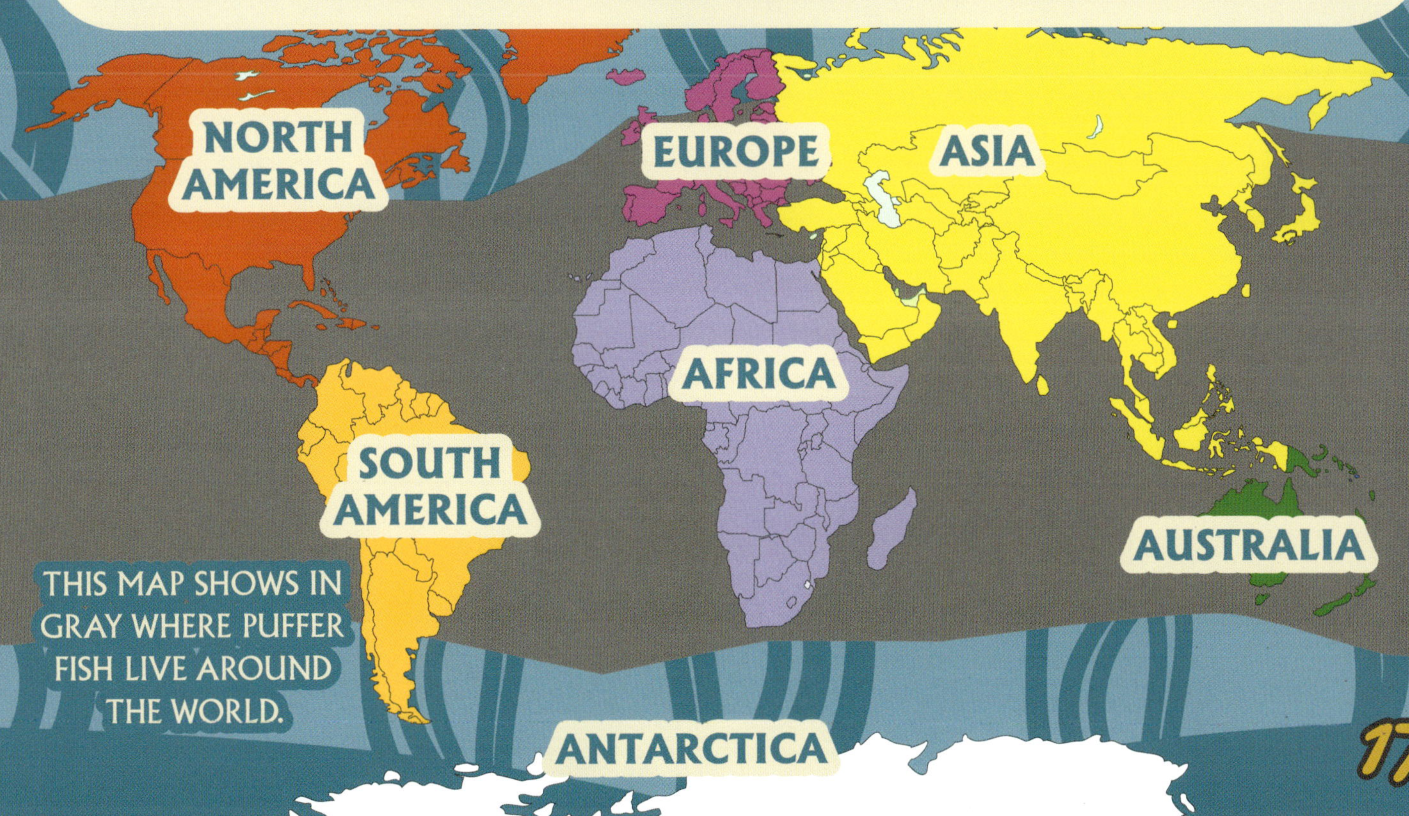

THIS MAP SHOWS IN GRAY WHERE PUFFER FISH LIVE AROUND THE WORLD.

PUFFER POISON

Puffer fish are poisonous. Poisonous animals make toxins that are harmful when touched or eaten. This is different from being venomous. Venomous creatures **inject** venom into their prey. They can use fangs, spines, or stingers to do this.

The toxic species of puffers can withstand about 500 to 1,000 times more tetrodotoxin in their bodies than nontoxic puffer fish or other fish.

PEOPLE WHO DON'T KNOW THAT MANY PUFFERS ARE POISONOUS MAY TRY TO PICK THEM UP. THE PUFFER'S SPIKES DO NOT INJECT ANY VENOM, BUT THE SKIN CAN BE COVERED WITH POISON. ALSO, THE PUFFER MAY BITE!

Puffer fish poison is called tetrodotoxin. Puffer fish don't make this toxin. Puffers become toxic by gathering and storing the toxin in their tissues. Scientists think puffers get this toxin from the food chain. Bacteria produce tetrodotoxin. These bacteria are eaten by creatures that are then eaten by other animals that are eaten by puffer fish.

TOXIC!

For many puffer species, the liver and ovaries are the most toxic body parts, followed by their intestines and skin. However, for some freshwater and brackish-water puffers, their skin is the most poisonous.

How do predators become poisoned by puffer fish? By eating them! When toxic puffers encounter enemies, they **excrete** tetrodotoxin from their skin.

LIZARD FISH EATING A PUFFER FISH

DWARF PUFFER FISH ARE A POPULAR CHOICE FOR HOME AQUARIUMS.

Tetrodotoxin makes puffers taste terrible, so predators often spit them out—fast! But the unlucky ones who eat poisonous puffers can die. Even though they are deadly, some people like to keep them as pets!

EFFECTS OF THE TOXIN

Tetrodotoxin is a scary toxin. The effects of the poison can start acting in minutes, and death usually happens within hours.

A person who has unknowingly eaten the fish might develop a headache or nausea. They might throw up. In more extreme cases, the victim will lose the ability to move parts of their body. This is because the toxin gradually shuts down the **central nervous system**. This will cause the victim to have trouble breathing and can lead to death.

Get the Facts!

For the victims who don't die, the experience is horrible. Victims are alert and awake the whole time the poison is in their system.

MANY ANIMALS TRY TO EAT PUFFER FISH, BUT VERY FEW CAN DO IT SUCCESSFULLY.

DARING DINING

There's no **antidote** to tetrodotoxin. If a victim is brought to the hospital on time, doctors will pump out their stomach. Victims also may be put on a machine to keep them breathing.

This doesn't stop some people from ordering puffer fish for dinner! In Japan, fugu chefs must train hard. They study for years to learn how to remove all the poisonous parts from the fish. Deaths from eating puffer fish in Japan have fallen since the 1980s. Each year, however, three to four people die from fugu poisoning.

Get the Facts!
Fugu chefs-in-training have to take a written exam. Then they have to prepare the puffer fish—and eat it! Many trainees fail the test.
EVEN IF THE PUFFER FISH IS PREPARED PROPERLY, DINERS MIGHT EXPERIENCE NUMBNESS IN THEIR LIPS AND TONGUE.
FUGU

USES FOR PUFFER POISON

Besides being an interesting food source, puffer fish have other valuable uses. Scientists are studying how their poison can be used to help people.

Get the Facts!

Tetrodotoxin gets its name from the animal order Tetraodontiformes. This group includes puffers, as well as triggerfish and sunfish. Not all Tetraodontiformes are poisonous.

Puffer poison may be used to treat long-term pain. A drug called morphine is often used as a painkiller to help people going through hard medical treatments. Researchers have come up with a painkiller based on tetrodotoxin. It is 3,000 times stronger than morphine! Also, the puffer poison medicine does not have the same tough side effects, like nausea, or feeling like you need to throw up.

THE RAREST PUFFER

Most kinds of puffers are doing well, but some species are vulnerable, or open to harm. Pollution and overfishing threaten their numbers. Losing their homes, or habitats, is another problem for some puffers. For example, *Tetraodon pustulatus* is only found in the Cross River in Nigeria and Cameroon. Scientists worry that oil exploration could be a danger to this species. So could other problems in Korup National Park, where this puffer lives.

Puffer fish can poison both people and animals. Still, people around the globe eat them! Would you ever try fugu?

REDLINE PUFFER FISH

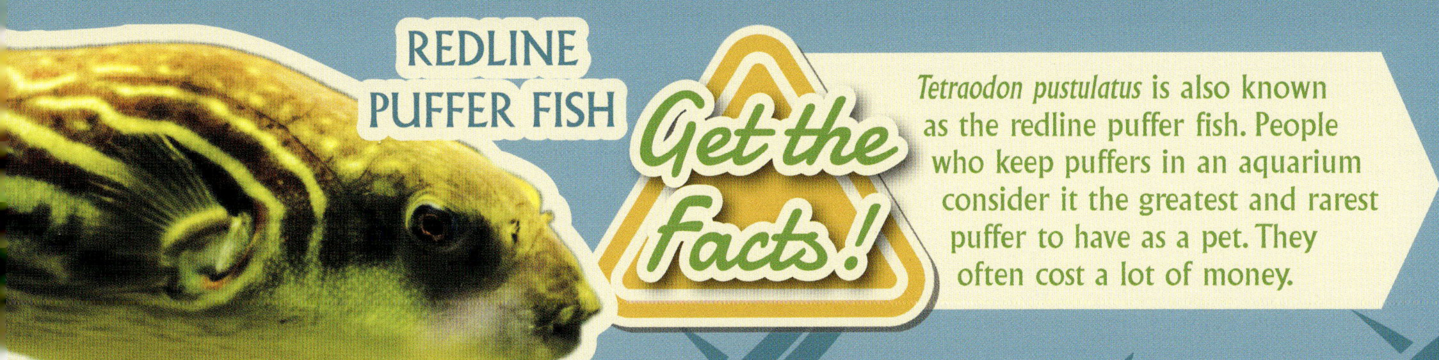

Get the Facts!

Tetraodon pustulatus is also known as the redline puffer fish. People who keep puffers in an aquarium consider it the greatest and rarest puffer to have as a pet. They often cost a lot of money.

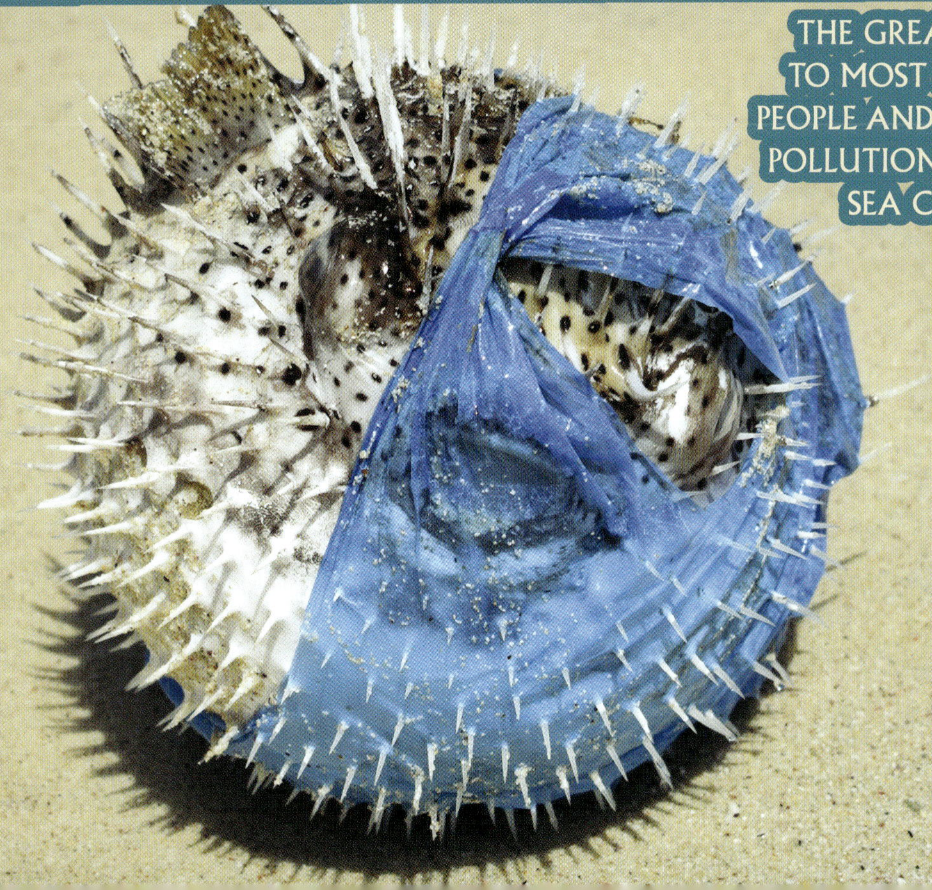

THE GREATEST THREAT TO MOST PUFFER FISH IS PEOPLE AND THEIR ACTIONS. POLLUTION HARMS MANY SEA CREATURES.

GLOSSARY

alga A plant or plantlike organism (like seaweed) that includes forms mainly growing in water. The plural form is algae.

antidote A drug that fixes the effects of a poison.

brackish Slightly salty, such as when river water and seawater are mixed.

central nervous system The part of the nervous system that includes the brain and spinal cord.

defense A way of guarding against attack.

excrete To pass waste matter from the body or from an organ in the body.

immune To have protection from something.

inflate To swell up with a gas or a liquid.

inject To force a liquid into a living thing using a needle, stinger, or fangs.

prey An animal that is hunted or killed by another animal for food.

tropical A region that never gets frost and that has temperatures warm enough for plants to grow all year long.

FOR MORE INFORMATION

Books

Bassier, Emma. *Blowfish.* Independently published, 2019.

Shea, Therese. *Pufferfish.* New York, NY: Gareth Stevens, 2017.

Websites

11 Amazing Puffer Fish Facts for Kids
www.growkido.com/puffer-fish-facts-for-kids/
Read some fascinating facts about puffers!

Pufferfish
kids.nationalgeographic.com/animals/fish/facts/pufferfish
Learn even more about puffers at this detailed website. Includes images and a video of a puffer puffing up!

INDEX